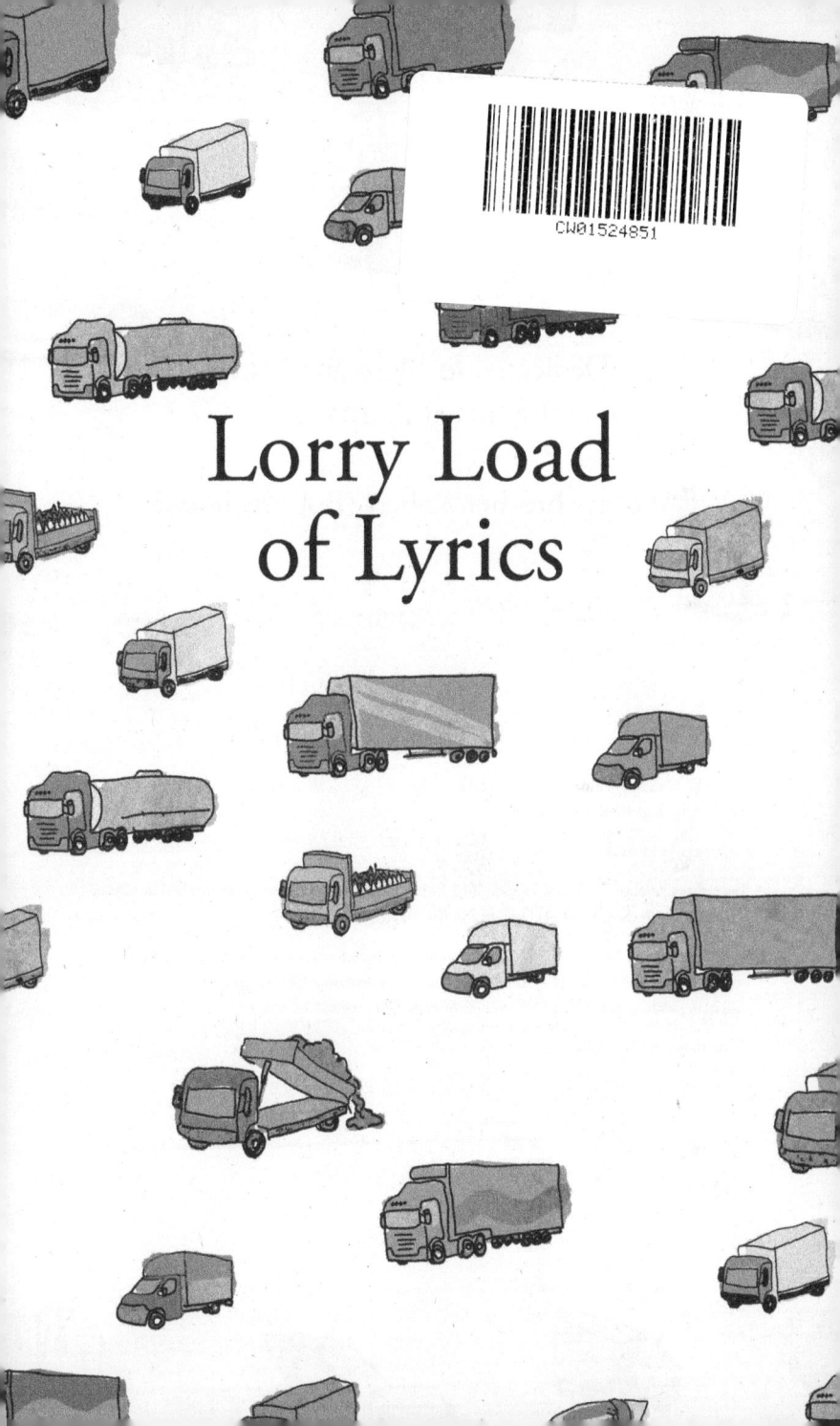

Lorry Load of Lyrics

Dedicated to Mom and Dad,
Brenda and my kids

And my brother Robert (Bob the book)

Crumps Barn Studio
No.2 The Waterloo, Cirencester GL7 2PZ
www.crumpsbarnstudio.co.uk

Copyright © Michael J Evans 2023

The right of Michael J Evans to be identified as the author of this work has been asserted by him in accordance with the Copyright, Designs and Patents Act 1988.

All rights reserved. No part of this publication may be reproduced, stored in a retrieval system, or transmitted in any form or by any means, electronic, mechanical, photocopying, recording or otherwise, without the prior permission of the copyright owner.

Cover design by Lorna Gray

Printed in the UK by Severn, Gloucester on responsibly sourced paper

ISBN 978-1-915067-37-1

Lorry Load of Lyrics

MICHAEL J EVANS

Collected Poems

Crumps Barn Studio

A NOTE FROM THE AUTHOR

I was born in February 1954 in Birmingham, the second son of Bill and Veronica Evans. My earliest awareness of the world about me was watching *Bill and Ben the Flowerpot Men* and hearing about Sputnik being launched. I had a typical upbringing for that time, playing in the streets, fetching coal in an old pram from the gasworks and living in a back-to-back house in Aston.

Dad was a lorry driver and whenever I could, I would bunk off school to join him on the road. I knew then what I'd like to be, even to the point of joining the army at 18 just to get my HGV licence early.

After the army I got married and we had four lovely children but we split and I moved out. It was then I decided to buy an old Motorhome THE BLUNDERBUS. To give the kids a sense of adventure, we would travel the country at weekends stopping at new places, then back to work on Monday as a lorry driver.

So that was my chosen career and I can say it served me well. During my work breaks I liked to write poetry. Being on my own in the truck, I based it on my observations in life and personal stories. I will share them with you. I hope you enjoy my words. Thank you.

LORRY LOAD OF LYRICS

CHILDHOOD HOLIDAYS

I loved my childhood holidays in Towyn
    near to Rhyl,
Long Abergele road we'd walk until we had
    our fill.
Or maybe on an open bus, the wind would
    cut our faces.
The sand and salt would blow about and
    get in awkward places.
Or if we wanted, and dad could find the money,
We would hire a tandem bicycle – now that was
    really funny.
We'd dash amongst the caravans until
    our time was up,
Then go for a game of bingo at a camp they
    called Winkup.
In the day we'd search for winkles upon the
    rocky shores,
In the evening we would cook them and eat
    them with crabs' claws.
Oh yes I loved my holidays in Towyn
    near to Rhyl,
And although I am now sixty-six
    I love to go there still.

FRIENDSHIP

Jeanie was a girl that I much admired,
But as romance goes nothing much transpired,
It was purely friendship as you can see
I was friend to her she was friend to me.

Playing in the park on a summer's day,
I would pull her hair and she'd run away,
Now that we are grown we live separate lives,
Mine in Birmingham hers down in St Ives.

Maybe we'll meet up once again some day,
I'll look in her eyes, won't know what to say,
Jeanie was a girl that I much admired,
But as romance goes nothing much transpired.

MY ANGEL

The fourth of September 1997
Is a date I will remember cause my mom
	went up to heaven.
She stood at the gates there with some
	uncertainty, saying,
"Can somebody guide me to where
	I'm supposed to be?"

Then an Angel stepped forward saying,
	"I know you my friend,
Shall we walk along together and then to
	God's house ascend?"

Mom said, "I m a little nervous but you
	calm me with your manner,
By the way my angel what's your name?"
"Well Vera it's Diana."

So off they went a step or two then a voice
	called from behind,
"Have you room for a tired old nun?"
"Of course Theresa, I'll also be your guide.

So off they went all hand in hand and
 walked on in the night,
Three angels then ascended into the
 heavenly light.

DESTINY

You are my destiny,
The compass points to you,
You can have the best of me,
When I get back to you
I wandered far away,
But always saw your smile,
My dreams were full of you,
I'll be there in a while.
So turn the light on in the kitchen,
Turn the light on in the hall
Turn your love on in the bedroom,
Where there's been no love at all,
Round the door tie yellow ribbons
Cause I'm coming home to you,
There can be no greater love dear,
Than the love I have for you.
The years have passed so slow
Since I went away,
You knew I had to go,
Although you begged me stay,

The battle's over now,
I've peace of mind at last,
Look to the future now,
Forget about the past.
We'll keep our love
And hold it fast.

HOPE

Don't look at me like that now, can't you
see the way I feel,
Don't you know I feel your sorrow now I
know that it's for real.

We've come all this way together, did you
think I'd leave you now,
When I said my love's forever, it was more
than just a vow,

Now we sit beside his bedside watching
monitors pulsate,
Will they find a cure to save him, will a
cure be found too late.

Hold my hand now will you partner, let us
say a little prayer,
Hoping someone else will hear us and
maybe answer out there,

Please stay with us till morning, hang onto
your little life,
You are everything we wanted, without
you we have no life

THE SPLIT

It's been a while since we last touched
Why did it come to this
I miss your smile,
I miss your laugh and of course,
I miss your kiss.
I thought that you were happy
And I guess I didn't see
All the pain and all the suffering
You were getting out of me
And now that I have lost your love
I'm all alone again
Baby come and rescue me
I don't want to go round again.
I call you on the telephone but you never
    call me back
Can't say I really blame you. You just don't want
    the flak.
So I guess we're really finished now,
And I've got to hide my pain
Go on by myself now, but
I don't want to go round again.

SPY

Who's that creeping round the car park?
Who's that slipping down the tree?
You're just making up excuses
You've been sleeping with the enemy.
Don't come here in all your perfume,
And your sexy lingerie
I'm not piggy in the middle,
You've been sleeping with the enemy
Do you think that I'm so stupid,
I can read between the lines
Someone else is playing cupid
While my love for you declines
I have seen your curtains moving
In the morning before three
Making love to 007
You've been sleeping with the enemy.

THE CLOCK

The clock with its hands keeps me in its sight,
The clock with its hands watches me all night,
And I look all around me. There's nobody there,
When you live on your own it's a
    saddening affair

So I wake up for work then I go through
    the day,
I come home in the evening with nothing to say,

Then I look at the box and I listen to sounds,
But my life is so empty it's getting me down,
And the clock on the wall pulses softly and low,
And I look to the window, but nowhere to go.

LONELINESS

All you lonely people,
Sitting in your sadness,
Kick that chair from under you
I have been where you're at
Like a neutered tom cat
Find it hard to heal the wounds.
Talking inward all the time
Makes you feel so lonely
Wrench those padlocks
From your mind
Talk to people boldly
Brighten personality
Neons deep inside you
Steer your love to other people
Then joy will come to you.

OUR ADVENTURES

Looking through my window, listening for
    the noise,
That will end my solitary bliss,
One daughter and three boys.

Once again they all come down for our
    regular weekend laugh,
The ex says send them all back clean,
But they rarely have time for a bath.

Off come the school shoes, and on go the boots,
Andrew checks road maps, for travelling routes
Liam loads food packs into the motor,
While Sarah writes down the washing up rota.

Richard is playing, we'll leave him alone,
Till we get to the mountains we all love to roam,
Then its stock up with chocolate to
    conquer the peak,
After which its back home,
Back to mom for a week.

LIAM

Thought my boy and me were getting there,
Thought we'd run that rocky road,
But I see that it's not over yet,
Keep on carrying my load,
It's been a burden hard to bear for his mother
    and for me
But somehow we are getting there,
And our son will then be free.
The sun shines in his eyes for us,
And for his uncle too,
There are times when he can make us cuss,
But we love him through and through,
Cause we know he's just a part of us
With problems. Just a few.
And for me, myself I'm hurting bad,
And I guess he's hurting too
Cause neither of us want this pain
But god what do we do.

SARAH JAYNE

Sweet little Ballerina apple of my eye,
Smiles and kisses, lots of love,
And occasionally shy.

Goldilocks and Vixen all one in the same,
That's my lovely daughter, my lovely Sarah Jayne.

Wild child of the playground, tougher
    than the boys,
Beats them all at boxing, then plays with
    Cindy toys,
Crying in the sunshine laughing in the rain,
That's my lovely daughter, my lovely Sarah Jayne.

Someday when you're grown up and I am old
    and grey,
Retain your sense of humour and never go astray.
And if you ever need me when life becomes
    a strain,
You know I'll always be there, my lovely
    Sarah Jayne.

EDALE

When I'm sad I think of Edale
When I'm mad I think of Edale
Then I'm glad I think of Edale
With its undulating hills.

Peace of mind is born in Edale
Soles of boots are worn in Edale
Grassy slopes adorn at Edale
There no dark satanic mills.

Though I am not part of Edale
I have left my heart in Edale
Soaring like a lark in Edale
Over valleys, brooks and stream

DRIVE SAFETLY

Don't drive fast in heavy rain
Or your car will lift
And you'll aquaplane
You'll slide like a hovercraft
With no control
Then you'll hit the barriers
Or maybe you'll roll
See the tyres can't cope
With that volume of water
But you bought them from new
And you think that they oughta
Well believe it from me
That it ain't no done deal
Keep your speed down for safety
And hands on the wheel.

DESPERATION

I was on the M6 motorway, driving down
    the middle,
A sudden urge came over me to pull up
    for a piddle.
Alas there were no services to satisfy my needs,
So I pulled up at the roadside and went amongst
    the reeds.
The flow seemed like forever but then
    finally it stopped,
As I sorted my attire it was then that I was
    copped.
"A problem with the car sir?"
No officer not the Cosworth
"Don't pull up unless you have to"
I said – me? More than me jobs worth,
He cautioned me for stopping. I said,
    sorry can't you see,
I was on a long, long journey and I needed
    for to pee
"Please don't get me wrong sir, I had to
    check you out.
You were driving so erratically and swaying
    all about."

OK so can I go now? He said, "Yes you can,
   my friend"
So I got back on the motorway, and journeyed
   to the end.

JOURNEY BY STEAM

Onward, onward, ever surging,
Forward on the shining track,
Jumping joints and points and junctions,
Steaming down to Crewe and back.
Wooden coaches shake and rattle,
Speeding through the station now,
People on the station's platform
Swaying from the slipstream's bow.
Stick your head out of the window,
And your face collects the soot,
Smell the burnt coal's rich aroma,
Duck in quick! the lineman's hut.
Hark! the mighty engine whistles,
Pulling us relentlessly,
Is it tunnel? Is it crossing? Is it station?
Wait and see.

CITY UNDER SIEGE

City under Siege,
Wipe back that tear for the enemy is near.
City under Siege,
Sharpen your blade, there's a price that
    must be paid.

They can't keep us holed up like rats forever,
You know and I know we'll have to box clever,
City under Siege,
We won't be here forever,
City under Siege.

City under Siege,
Reinforce defences to stop them
    coming through.
City under Siege,
Lay down your traps, the one they
    want is you.

The people are holding the place together,
Will they defeat us, the people say never.
City under Siege,
We'll have to box clever
City under Siege.

BLACK COUNTRY LAD

I cut me teeth on a piece of steel,
Them were heavy times
And I got me a job in Brierley Hill,
Them were heavy times,
I left school just afore the war,
And I still couldn't count over twenty-four
But I knew what a foundry man was for
Them were heavy times
Yes them were heavy times,
Forging chains for the ships of the world
Then out for a pint and a kiss from a girl
And we'd settle in the pub and hear the
    tales unfurl
Them were heavy times.
The old uns said they had it bad
Them were heavy times
But I worked just as hard as my dear old dad
Them were heavy times
I could drink ten pints of Banks's ale
Eat groaty pudding and live to tell the tale
I'm a drunk but proud black country male
Them were heavy times
Yes them were heavy times

Forging chains for the ships of the world
Then out for a pint and a kiss from a girl
And we'd settle in the pub and hear the
 tales unfurl
Them were heavy times, yes them were
 heavy times.

THE BARGEE

Don't ever fall into the cut, it's not
    the place to be,
Don't ever fall into the cut for it was
    the death of me.

I had a boat and butty, working the oxford line
I'd take a load from Birmingham onto the
    Thames sometime.
all shiny plates and cutlery, badges and
    buttons too.
Cause them folks up in Birmingham are the best
    at what they do.

Anyhow, to cut it short I lost my grip
    at Napton,
I fell off the boat into the lock, crushed my ribs
    and cracked them.
The wife and kiddies tried in vain to lift my
    floating body,
But all the life had gone from me and I lay there
    cold and soggy.

So don't ever fall into the cut it's not
 the place to be,
Don't ever fall into the cut for it was
 the death o me.

BULLY

Gertcha! That's what my old man said
Every time he hit me bout the head,
Everytime he whacked me with he's belt,
I felt it all, but nought he felt.
My momma cowered and did her deeds
And catered to his constant needs,
She'd fetch his beer and wash our clothes
And try to dodge his wicked blows.
Then one day aged twenty-four, I reared up
Could take no more,
I beat down on his empty shell
And sent him packing off to hell.

SEX SLAVES

We came to this country with innocent eyes,
Their open arms met us but we were not wise.
They plied us with liquor and promises bold,
We never suspected our bodies were sold.

And then came the punters some young and
    some old,
Their money gave hope as together we rolled
Neath the sheets all so silken and stained
    by the use
There may have been love mostly sexual abuse.

But we held our heads high as we offered
    us up,
To the men who were lonely and cried in
    their cup,
To the fresh faced young virgin who
    blushed when him kissed.
To the half cut old sailor who sleeps
    when he's pissed.

And when sleep comes at last what a
 beautiful thing,
To dream of Sumatra, the place I lived in.

RUDE BOY

Rude boy Winston came to town,
Lookin up what's goin down,
Came to Brixton wide of eye,
They knew he was kinda shy,
Took him in and gave him love,
Gave him coke and then the shove,
Rude boy broken, in the gutter,
Female eyes began to flutter,
Brought him back into her flat,
He ran out ten seconds flat

Rude boy what you come here for?
London's bad, you know the score,
Best advice to keep away
Be rude boy another day,
Crack and coke it ain't your scene,
You been brought up good and clean,
Say your prayers most every day,
Saints and sinners keep away,

Rude boy you go back to school,
Let some other be the fool,
Be a symbol for the brothers,
Make them be proud for their mothers.

SONG MAN

He was just a song man not a strong man,
and his songs went deep to your soul.
He would play you some new songs
And he'd play you some old.
He was just a song man not a strong man,
And his songs went straight to your heart
But he's gone now and I see how
His own heart aches tore him apart.
They tore him apart.
There was life and colour in his music,
Hiding all the pain felt inside
He said he was happy in today's world
Now we know that he lied
We know that he cried.

TUNE

I wrote a tune on my piano,
    my fingers stroke the ivories,
And if you hear some notes that please you
Or some that tease you, I will be pleased.
A party tune or some sweet melody,
It's up to you just what I play.
And if you spend some time and stay with me
And even play with me,
We'll chase those blues away.
My tune is finally completed,
The softness of the tone is sweet,
The melody is now repeated,
    the night's retreated
Until the next time that we meet.

THE AGEING ROCKSTAR

Elvis can I join you in that mansion in the sky?
I have played so many concerts that I think
    I'm about to die.
I'm rounding off near eighty,
But they want me to stay cool,
I keep churning out new records
And they buy them still, the fools
I'd just like to sit back and reflect upon my life,
Why I haven't got a future why I haven't
    got a wife.
I would love to hold my children, if I knew just
    who they were
But so many painted women and the drugs
    stopped that for sure

So that's the way it is king, was it like that just
　　　for you
Were you tired of the rat race?
Did you know your time was through?
Cause I'm feeling pretty low now and my
　　　fingers are so numb,
I can't do no finger picking and I find it
　　　hard to strum
My skintight jeans don't fit now, and I have to
　　　dye my hair
So spare a thought for this old Rock star,
　　　and welcome me up there

THE SALESMAN

I used to be a salesman no one kept my
 calling card,
I used to do the business but I found it
 very hard,
But now I am a pop star, fame and fortune are
 my friends
I sing at different nightspots and I hope it
 never ends.
When the curtain's up I wow them
With my songs and sexy moves,
Hard to think I was a salesman struggling
 selling shoes
A chauffeur-driven limo takes me where
 I wanna go
You see I am really busy,
Excuse me while I do this show.

BRENDA

Brenda, you are my light and joy,
My rainbow in the sky,
You cheer me up when things are bad,
You're the apple of my eye,
I love you always,
And fear ye not,
I'll never say bye-bye.

WRINKLIES

A chirpy little thing you are,
With so much energy,
I've got to watch my step with you,
Or you'll be the death of me,
I never expected to have such fun,
With a woman of senior years,
You shine a light into my life,
With all your thoughts and cares,
OK so we are wrinklies,
but not within our minds,
we have such joy as all the rest,
Brenda your love shines.

THE SOMME

Say a prayer for Johnny Allen,
We were there to fight the foe,
We were comrades in the trenches,
Till it was time to go.
Say a prayer for Johnny Allen,
Blood runs cold within his veins,
Eyes so glazed as if he's crying,
He no more shall feel the pains
Cry no more for Johnny Allen,
Or the gas will burn your eyes,
Barbed wire rips into your tunic,
Smells of death and swarms of flies,
Weep no more for Johnny Allen,
Can't say I knew him well,
We were brothers for a short while,
Shared some bread amidst the hell
Johnny Allen I may see you, on the other side,
Battles started bombs are falling,
Will you stay and be my guide?

THE TREE

Look out the window and what do I see?
A car, a camper, and a bloody great tree.
They say it's an eyesore but I disagree,
Cause me and the tree are buddies you see,
It waves to me daily as I go by,
It shades me from sunshine way up in the sky,
In autumn it covers the drive with its leaves,
It looks kind of sad then but it never grieves
Cause it's only sleeping till winter goes by
Till the summer warms up from the sun
    in the sky,
Then the leaves start to grow and the
    branches do sway
Well they're waving to me as I go on my way.

END IS NEAR

I dug a hole in my planet today,
And took its future life away,
I dug for diamond and for gold,
And other minerals to be sold.
I drained the oil from deep within,
And stole the land that nature's in,
And when I finally saw the light,
I began to feel sad and so contrite,
For I found that it was far too late,
For me and mankind to abate.

ZOMBIES

There's a smell, something's in the air,
Well you can run but you'll get nowhere,
Don't fight it. Stop! They're coming after you,
Remember Zombies, Zombies need love too.
Bolt the doors, get your crucifix,
They've got you now. Boy you're in a fix,
Can't say no more there's nothing I can do
Zombies, Zombies need love too,
And when one holds you in its embrace,
Smile at its decomposing face,
There's nothing else that you can do,
Just remember they need love too.
Terror dreams racing through you're mind,
Try to run but you're running blind,
Give into them, they have something for you,
Remember Zombies, Zombies need love too.